# MEGA TANKS

Written by Paul Stevenson

# CONTENTS

| | |
|---|---|
| Tank Warfare | 4 |
| History of the Tank | 6 |
| The Challenger 2 | 8 |
| The M1 Abrams Tank | 10 |
| Tank Crew | 12 |
| The Driver and Gunner | 14 |
| At Night | 16 |
| Inside the Tank | 18 |
| Tanks at Work | 20 |
| Prepared Position | 22 |
| Tank Under Attack | 24 |
| Tanks of the Future | 28 |
| Become a Crew Member | 30 |
| Glossary | 31 |
| Index | 32 |

First published in 2024 by
Hungry Tomato Ltd
F15, Old Bakery Studios,
Blewetts Wharf, Malpas Road,
Truro, Cornwall,
TR1 1QH, UK.

Copyright © 2024 Hungry Tomato Ltd

No part of this publication may be reproduced, stored in a retrieval system, or transmitted in any form or by any means, electronic, mechanical, photocopying, recording, or otherwise, without prior written permission of the copyright owner.

A CIP catalogue record for this book is available from the British Library.

ISBN 9781916598836
Printed in China

Discover more at
www.hungrytomato.com

All words in **BOLD** can be found in the glossary.

# TANK WARFARE

Tank warfare is when an army uses **armoured** fighting vehicles against an enemy. Tanks attack the enemy head-on.

Tanks can fire on the enemy using a large-**calibre** gun and machine guns.

Heavy armour and good **mobility** give the tank protection.

Machine guns can turn 360 degrees to face the enemy

Turret

Caterpillar tread

Wide, thick tracks allow the tank to move fast across rough ground. They spread out the tank's weight. This means the tank is less likely to get stuck in soft ground, mud or snow.

**Large-calibre gun**

**Heavily armoured hull**

**Tough treads which do not puncture or tear, unlike tyres**

# HISTORY OF THE TANK

**During World War One, trench warfare was used, which meant that soldiers hid in trenches.**

They had to climb out of the trenches and run across open land to attack the enemy. This was called "going over the top".

Trench

Soldiers "going over the top"

Very little **territory** was captured using trench warfare. Millions of soldiers were killed.

The British Army wanted a new way to capture enemy territory. They decided that they needed a vehicle which could move over rough ground, fire at the enemy and protect the soldiers inside.

British Mark A Whippet tank in 1918

It went from an idea to the battlefields of World War One in less than three years.

The invention of the tank helped to end the war.

# THE CHALLENGER 2

**The FV4034 Challenger 2 is the British Army's main battle tank. It is very heavily armoured.**

The British Army has around 227 Challenger 2 tanks, though many are not in a position to be deployed.

Here, a Challenger 2 tank rolls off a transport ship in Kuwait.

Cover to protect the gun in transit

Fuel drum

Rear of tank

## FV4034 Challenger 2

**Cost:** Over £4,000,000 British pounds each
**Top Speed:** Up to 37 miles per hour (mph)
**Weight:** 75 tonnes
**Crew:** 4 crew members

Transport ship

# THE M1 ABRAMS TANK

**The M1 Abrams is the U.S. Army's go-to tank.**

The Abrams is nicknamed "Whispering Death" because its engine makes so little noise. The enemy can't hear it coming!

The top panels of the tank blow outward if it is hit by a HEAT (High-Explosive Anti-Tank) missile. This stops pieces of the damaged tank from hurting the crew inside.

Turret

Engine

Chobham armour

Top panels

# M1 ABRAMS TANK

**Cost:** Over £8,000,000 British pounds each
**Top Speed:** Up to 46 miles per hour (mph)
**Weight:** Over 62 tonnes
**Crew:** 4 crew members

# TANK CREW

**Tank crews usually have 4 members: a commander, operator, driver, and gunner.**

The full tank takes over 1,900 litres of fuel, giving it an operational range of up to 124 miles.

Gunner

Driver

Each member of the crew has a job to do. However, they are trained to do one another's jobs, too. This is important in case a crew member is unwell or injured.

The operator works the **radio** and loads rounds into the guns. In bigger tanks, these jobs are done by two different people: the radio operator and the loader.

Commander

Operator

# THE DRIVER AND GUNNER

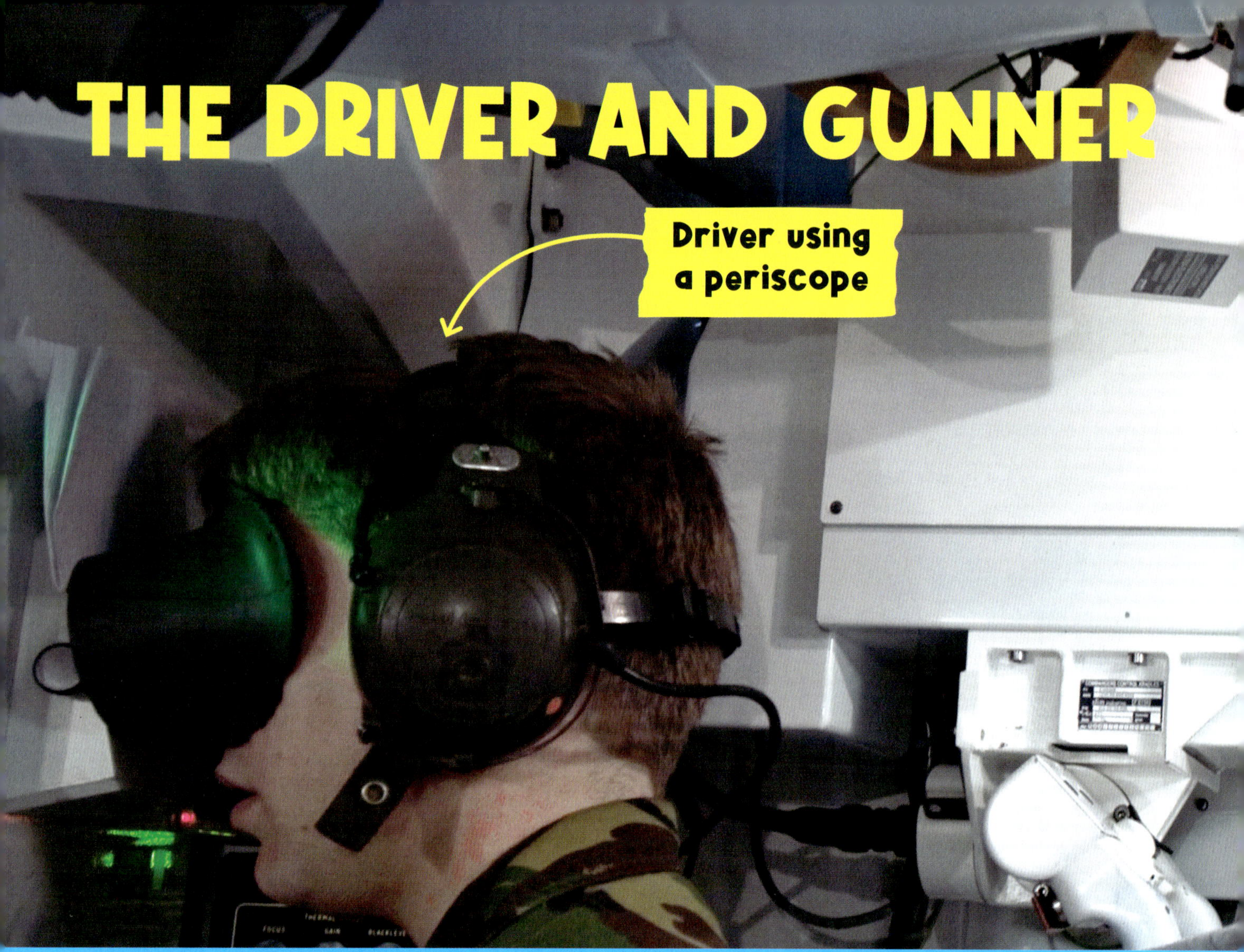

**Driver using a periscope**

## THE DRIVER:

**The tank driver sits in the front of the hull, under the main gun, with the periscope on.**

To fit in the small space, he has to lean back.

The driver steers the tank using a motorcycle-style handlebar. The handlebar has a throttle which the driver twists to make the tank move faster.

Gunner

## THE GUNNER:

The gunner controls the direction and angle of the **turret** and main gun.

The gunner pinpoints targets using **laser range finders** to find out how far away the target is.

15

# AT NIGHT

At night, the crew uses thermal night-vision to see outside of the tank.

This equipment allows soldiers to see enemy tanks or soldiers without using any lights.

It's vital for the crew of a tank to have 24-hour sight, night and day in battle.

# INSIDE THE TANK

**There's not much room inside a tank. Four people must share the space, sometimes for days at a time!**

If the tank is in a desert, it can get very hot inside during the day. At night, it can be freezing cold.

When on duty, a tank crew lives, sleeps, and works in the same clothes for days.

A member of the crew taking a break

There's no place to wash in the tank and it's too risky to wash outside. The smell of soap can carry on the wind and give your **position** away to the enemy!

The toilet is a hole in the ground dug out with a shovel. When on duty, the crew eats MRE rations – Meals Ready to Eat. These dry, **vacuum-packed** meals can be boring and tasteless.

Thankfully, the crew gets chocolate bars and sweets in their rations!

# TANKS AT WORK

**Tanks hardly ever work alone. Normally, they work in a platoon of five.**

Two tanks might **advance** while the other three tanks protect them from behind.

Then, the two tanks in front stop. The three tanks at the back then advance while the front two cover them.

Tank platoon

A tank's turret can turn a full circle – 360 degrees. The tank can shoot backwards while moving forwards.

Each tank in a platoon can have guns pointing in different directions.

This gives the platoon protection from all sides. The tanks can advance without stopping or slowing down.

Turret

# PREPARED POSITION

When defending territory, tanks stay in something called a "prepared position". This helps to protect the tank.

This tank is covered in leaves and green nets as camouflage

A prepared position can be a hole or behind a hill. Only the turret shows. This means that the crew can fire on the enemy, but most of their own tank is protected.

Sometimes, tanks use trees as **camouflage.**

Tank commanders use the IVIS (Inter-Vehicle Information System) to keep in touch with other tanks.

They also send maps and share information about the enemy. They use radio signals that are in code.

# TANK UNDER ATTACK

**Challenger 2 is one of the most heavily armoured tanks in the world.**

While on duty in Iraq in 2003, a Challenger came under attack. The enemy used machine guns and RPGs (Rocket-Propelled Grenades).

The tank tried to reverse, but it fell into a ditch. The tank's tracks came off!

## THE CHALLENGER WAS **TRAPPED!**

The tank was hit by an **anti-tank missile** and 14 RPGs at close range. Luckily all the crew survived!

The crew stayed safely in the tank until they were rescued. The tank's tracks were repaired, and it was back to work just six hours later.

**The easiest way to stop a tank is to hit the tracks. Once these are damaged, the tank can't move. This is called a "mobility kill".**

The tracks and wheels of a tank are outside the armoured hull.

When a tank goes over a hill, the enemy can fire at its underside.

The armour on the underside of a tank is not as thick as on the hull.

Tanks are easier to damage in cities; enemies can fire on a tank from the top of a tall building, because the top of the tank is not as heavily armoured as the hull. This makes it easier to damage!

# TANKS OF THE FUTURE

In modern warfare, smaller, faster armoured vehicles are often more useful to armies than the main battle tanks.

Because of this, a new range of light armoured tanks have been built.

Scimitar

Scimitars and Spartans are small with top speeds of 50 mph. They can get close to an enemy and report back on enemy positions and numbers. They are harder to spot than the larger Challengers and Abrams.

## HOWEVER, IF THEY ARE SPOTTED, THEY HAVE THE FIREPOWER TO FIGHT BACK!

Spartan tanks being transported by Chinook helicopters

# BECOME A CREW MEMBER

## TRAINEE REQUIREMENTS:

- To join the army as a tank crew member, you must be over 17.
- No special qualifications are needed.
- You must be fit.
- You must have good eyesight because you will need to read maps, find targets and drive tanks around obstacles.
- You must be a person who can work in a small, tight space for days at a time.

## BASIC TRAINING:

You will go on basic training to teach you military skills, such as handling and firing a gun. You will learn to live and work in the open air. You will build up your fitness levels.

**If you pass basic training, you can then train to become a tank crew member.**

# GLOSSARY

**advance** - to move forwards.

**anti-tank missile** - powerful missiles made to destroy armoured vehicles in battle.

**armoured** - something that has a protective covering of metal.

**calibre** - the measurement of the diameter of a gun's barrel.

**camouflage** - something, such as leaves or a colour, which helps a tank blend in to its background.

**Chobham armour** - an extra hard armour made up of layers of ceramic in a metal frame. Tanks made with Chobham armour are super resistant to HEAT weapons.

**laser range finder** - a piece of equipment that uses a laser to work out the distance of an enemy object.

**mobility** - the ability to move easily.

**periscope** - a piece of equipment that allows a soldier to see outside the tank while staying hidden inside.

**platoon** - a group of tanks or soldiers.

**position** - the place where an army or group of soldiers is hiding or waiting to attack.

**radio** - a piece of equipment that allows soldiers in different tanks to talk to each other, and to their headquarters. They can also use the radio to send data.

**territory** - an area of land that belongs to a country or that has been captured by an army.

**turret** - the top section of a tank where the guns are mounted. The turret can turn in a full circle to point the guns in any direction.

**vacuum-packed** - a way of packaging food so that there is no air around it. This stops the food from going bad.

# INDEX

**A**
Abrams (M1) tanks 10-11
armour 4-5, 8, 10, 24-25, 26-27, 28, 31

**B**
British Army 7, 8

**C**
camouflage 22-23, 31
Challenger 2 (FV4034) tanks 8-9, 24-25
Chobham armour 10, 31
commanders 12-13, 23
cost (of tanks) 9, 11
crews 9, 11, 12-13, 18-19, 25, 30

**D**
drivers 12, 14

**G**
gunners 12, 14-15
guns 4-5, 8, 12-13, 15, 21, 24

**H**
history of the tank 6-7

**I**
invention of the tank 7
IVIS (Inter-Vehicle Information system) 23

**L**
laser-range-finders 15, 31

**M**
M1 Abrams (see: Abrams)
mobility (of tanks) 4-5, 31
MRE (Meals Ready to Eat) 19

**O**
operators 12-13

**P**
periscopes 14, 31
prepared positions 22-23

**S**
Scimitars 28

soldiers (protection of) 7, 16, 31
Spartans 28-29
speed (of tanks) 9, 11, 28

**T**
tank life 18-19
tank platoons 20-21
tank warfare 4-5
tanks 4-5, 7, 8-9, 10-11, 20-21, 26-27, 28-29
thermal night-vision 16
tracks 5, 24-25, 26
training 13, 30
trench warfare 6-7
turrets 4, 10, 12, 15, 21, 23, 27, 31

**U**
U.S. Army 10

**W**
weight (of tanks) 9, 11
Whippet tanks 7
World War One 6-7

Picture credits:
(t=top; b=bottom; c=centre; l=left; r=right):
Shutterstock: CombatcameraUK 18bl; Karasev Viktor 27b; Karlis Dambrans 10-11; Martin Hibberd 28; Mike Mareen 2-3; Robert Sarnowski 1, 12-13; Saeediex 24b; StunningArt 25; Vladimir Kovalchuk 18-19; Wavebreakmedia 30. AFP/Getty Images: 15B, 22-23, 31b. Getty Images: 16-17. Jack Sullivan / Alamy: 21b, 26c. Leif Skoogfors/CORBIS: 14b. Photograph by: Richard Ellis; © Crown Copyright/MOD, image from www.photos.mod.uk. Reproduced with the permission of the Controller of Her Majesty's Stationery Office: 29. Photograph by: Sergeant Paul Brownbridge; © Crown Copyright/MOD, image from www.photos.mod.uk. Reproduced with the permission of the Controller of Her Majesty's Stationery Office: 14-15t. Photograph by: Sgt B Gamble; © Crown Copyright/MOD, image from www.photos.mod.uk. Reproduced with the permission of the Controller of Her Majesty's Stationery Office: 20. Photograph by: WO2 Giles Penfound; © Crown Copyright/MOD, image from www.photos.mod.uk. Reproduced with the permission of the Controller of Her Majesty's Stationery Office: 8-9. Rex Features: 4-5. Time & Life Pictures/Getty Images: 6.

Every effort has been made to trace the copyright holders, and we apologise in advance for any unintentional omissions. We would be pleased to insert the appropriate acknowledgements in any subsequent edition of this publication.